AN EARTHTOPIA BOOK

100 WAYS YOU CAN HELP SAVE THE PLANET

MICHAEL JOSEPH

PENGUIN MJ Est. 1935

AN EARTHTOPIA BOOK

100 WAYS YOU CAN HELP SAVE THE PLANET

DESIGN & ILLUSTRATION BY

SAMUEL B. THORNE

PENGUIN MICHAEL JOSEPH

UK | USA | Canada | Ireland | Australia
India | New Zealand | South Africa

Penguin Michael Joseph is part of the Penguin Random House group of companies
whose addresses can be found at global.penguinrandomhouse.com

First published by Penguin Michael Joseph 2024
001

Printed and bound in Great Britain by Clays Ltd, Elcograf S.p.A.

The authorized representative in the EEA is Penguin Random House Ireland,
Morrison Chambers, 32 Nassau Street, Dublin D02 YH68

A CIP catalogue record for this book is available from the British Library

ISBN: 978–0–241–64306–8

www.greenpenguin.co.uk

THANK YOU TO EVERYONE WHO HAS SUPPORTED EARTHTOPIA OVER THE YEARS:

ROB GREENFIELD • SAMUEL B. THORNE • DOMINIC COOK
ALICE REGESTER • SEBASTIAN UNDERHILL • MIKE BROWN
ANNA VINCENT • AMBER CARR • BECKY BURCHELL
LYDIA THURLOW • MILO FISHER • DUPÉ CRAIG
SEAN SPECZYK • TAMARA PENFOLD • STUART LARKIN
TATSU ISHIKAWA • JAMES POLETTI • NAOMI BARKER
GEORGE HAYLEY • SONIA BENDJOUDI • SARAH GIBSON
JACK FERRIS • LAURA GRANT • JAMES MONTAGUE
MARTIN FEARN • ELISAH VAN ALLEN • MEERA TRIVEDI
FRAN ARKELL • DILLON BRANNICK • DANIEL JAMIE WILLIAMS
ADRIANNA KRZOSKA • HARRY ROWETT • ELLIE HOLLAND
ANNA PERESTRELLO • CHARLIE DAVIS • FRED STANLEY
INJILA BAIG • ROSIE LEADER

Introduction

Earthtopia started as a community on TikTok. A place that offers hope amongst the doom & gloom of the climate crisis. Somewhere for people to come together and discover the actions and solutions that protect and restore the planet.

The daily content has inspired millions of people to make a change, to appreciate the Earth more and to celebrate the good news – because there's plenty of it.

And just like on TikTok, in this book you'll find straightforward and easily actionable tips that you can introduce into your everyday lives and know you're playing your part.

Whether that's choosing a more sustainable product, learning about activism or improving your local community.

But don't just stop there. Share any new found knowledge with friends, parents and, most importantly, remain hopeful.

Because the Earth deserves fighting for.

Join our online community!
Search Earthtopia on TikTok & Instagram.

Chapter 1

Getting Started

Don't give up hope!

Wildfires, droughts, floods, heatwaves, melting ice – the climate crisis is clear, but what can you do to stop it?

First things first, don't give up hope.

Much of the media and climate doomers will try to convince you that we've ruined the planet beyond a point of no return, but we're not there (yet).

There are tonnes of actions you can take that will have a positive impact on protecting and restoring our home.

So read on for how to save the planet, one step at a time.

It's okay to worry

Eco-anxiety is the worry you experience
when thinking about the climate crisis.

It can make you feel helpless, angry or sad.
But the first step to combatting it is to accept it
will be part of your journey as you fight for the planet.

And you're not alone. Studies suggest 60% of young people
wordwide are 'worried' or 'extremely worried' about the
climate crisis.

The tips in this book will help you take back power
and have a positive impact on the planet.

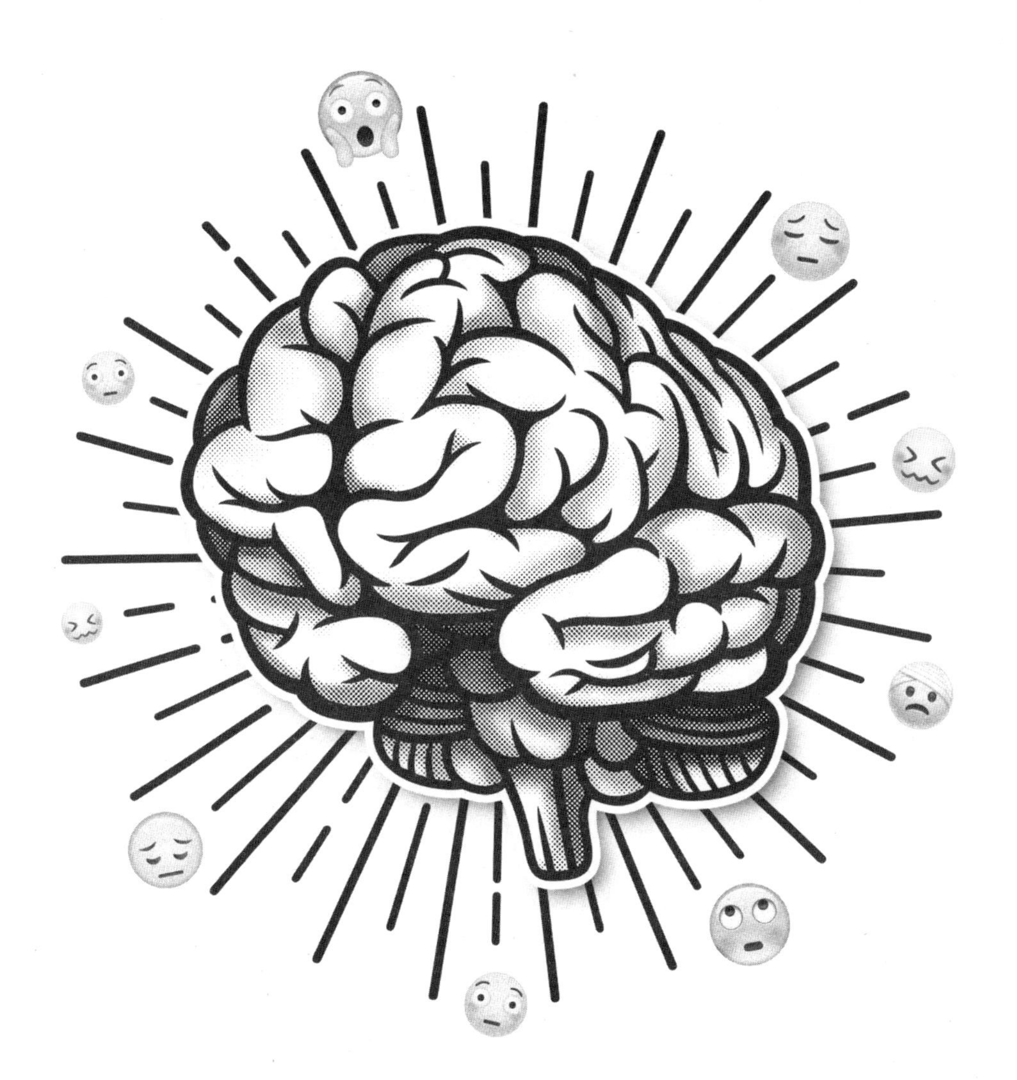

Learn about climate history

You know how your Dad lets out more farts in the house than everyone else? Well, the same can be said for different countries when it comes to greenhouse gas emissions.

The wealthiest countries have historically been responsible for the highest emissions, with the US and China being the worst offenders (the UK is fifth in that list), due in part to their huge industrial power, which was often founded on the colonisation and exploitation of poorer countries.

Whilst poorer countries statistically contribute lower emissions, they are disproportionally impacted by climate change and also have fewer resources when it comes to fighting against it.

It's important that we educate ourselves about the complex story behind climate inequality, and do our bit to make sure history doesn't repeat itself so that we ALL have a future to look forward to, regardless of where we happened to be born.

Don't try to save the planet on your own

Fighting to restore the planet can make us feel like superheroes (cape optional), on a lonely mission to save the world. But sometimes even superheroes need a sidekick or two.

Doing something as big as taking on the climate crisis all on your own is a surefire way to feel burnt out, frustrated and helpless. But when we join a movement or mobilise our friends, family and like-minded people, we have more power to influence our schools, communities, businesses and even governments to fight for the planet.

Luckily, we've gathered some handy tips on what movements you can join, how you can support those around you to make changes and also use the power of politics to create a positive impact. Just keep reading.

Rebalance the climate scales

So you've just found out the country you live in has a shady climate past. But your lifestyle isn't that bad for the planet, right?

Well... there's 3 billion people in the Global South who use less energy on an annual basis than a standard American fridge.

Whilst the richest 10% in the world (who are mainly in the Global North) are responsible for half of global CO2 emissions.

Consider the impact of your daily life, including travel, shopping habits, holidays and dinner plans:
Do you eat meat at least once a day?
Do you always drive to the supermarket?
Are you taking 2+ flights abroad each year?

If the answer is 'yes' to any of these, it's important to acknowledge your lifestyle, which is most likely the norm in your country, is not helping the situation. On the upside, there are lots of actions in the following chapters that will help you reduce your impact and inspire you to change your country's habits too.

Be an ally to Indigenous Peoples

Indigenous peoples are fewer than 5% of the global population, but protect 80% of global biodiversity.

They're teeming with knowledge on how to live in harmony with nature and reap the benefits without damaging the planet. But their ways of life are under threat from damaging activities like oil pipelines, beef ranches and mining.

To keep their knowledge alive, read books by Indigenous authors, and educate yourself about Indigenous cultures and planet-saving practices. Follow and share content from Indigenous leaders on social media and support the call for giving back stolen land.

It's good to talk

Now that you're learning to frame your feelings about the climate crisis and have explored some of the history and inequality associated with it, it's time to get talking.

Talk to your friends and family, see how they're reacting to the latest climate news, and tell them how you're feeling too.

Conversation sparks solutions, and sharing our knowledge – as well as our feelings – is one of the best tools we have in the fight to save the planet.

Gift this book

If you've already bought this book then congratulations, you're a step ahead when it comes to saving the planet – nice one!

Once you've read it from cover to cover, you can share your knowledge (and get into the spirit of recycling!) by passing it on to friends and family so they can start living more sustainable lives as well.

Whether you're re-gifting or buying brand new, books make great presents, and this one helps the planet too!

Chapter 2

Fashion

Ditch fast fashion

We're firm believers of quality over quantity when it comes to clothes. Of course we wanna look good, but we don't need hundreds of different outfits.

Cheap, poorly made clothes that use synthetic materials are the backbone of the fast fashion industry, alongside the workers they mistreat and pay next to nothing to make them.

So give fast fashion the finger by buying from sustainable brands if you can. The price tag can be more expensive, but just remember, you're buying something that'll keep you looking fresh for a long, long time.

Get thrifty

Wanna know another reason to put a finger up to fast fashion?

The cheap, poorly made garments that are mass-produced mean we throw away an entire rubbish trucks' worth of clothes every second.

So, why not get thrifty at your local charity shop or second-hand store?

It's kinder to our planet, fun, affordable, and that feeling of finding that 10/10 garm you never knew you wanted? Nothing beats it.

Wear less water

You're probably wearing a lot of water right now.

To make one pair of jeans it can take up to 10,000 litres of water, which is 10 years' worth of drinking water for one person.

Cotton is a thirsty, chemical-heavy crop, and anything synthetically made, like nylon joggers or polyester socks, will release microfibres into the ocean throughout its life cycle and won't break down if it ends up in landfill.

Look for organic cotton, as it avoids hazardous pesticides, or consider clothes made from super sustainable natural materials like bamboo or hemp.

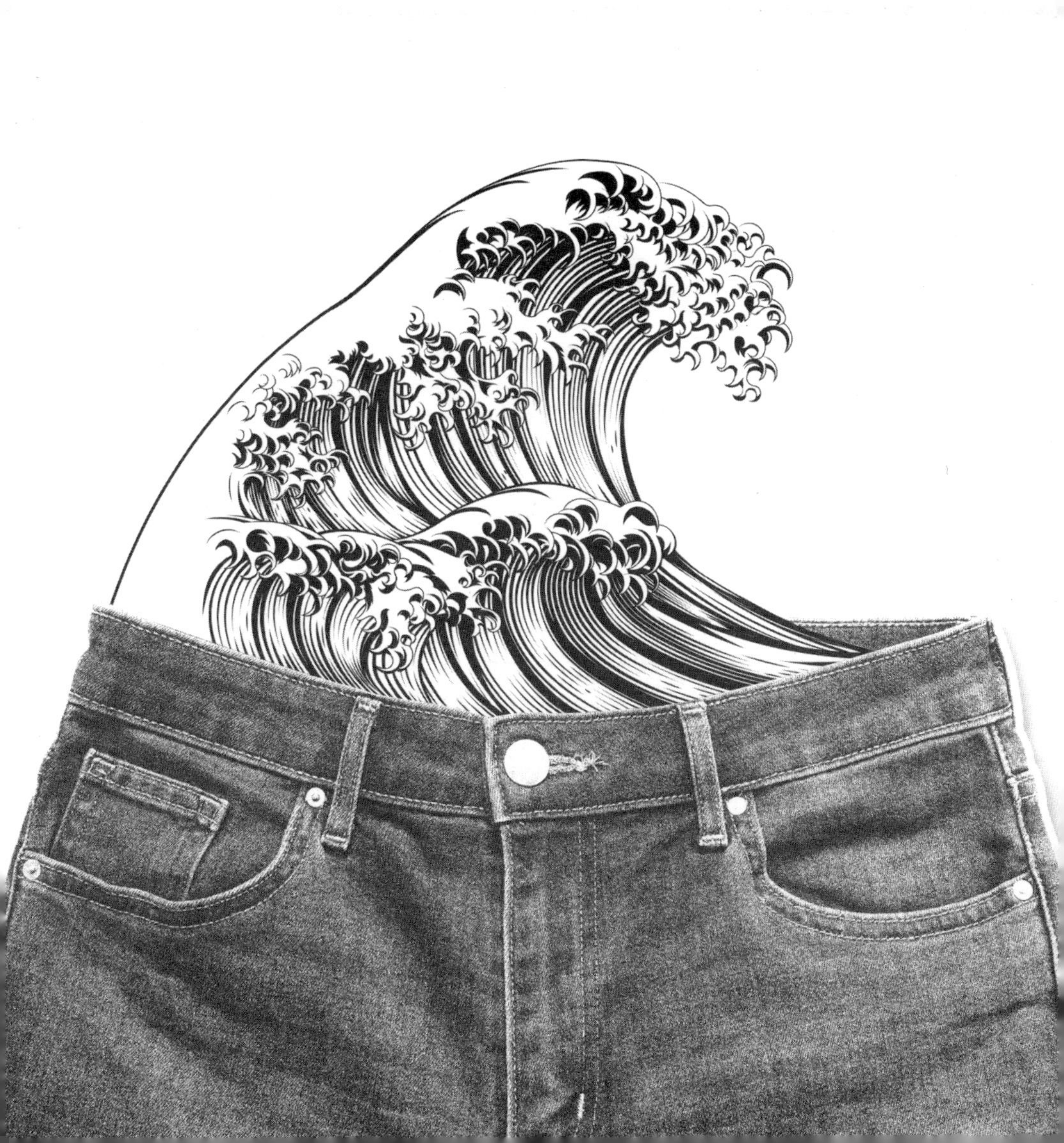

Do the smell test

Whilst we're on the topic of water... know what stinks? The 700,000 plastic microfibres released during the average clothes wash.

That means a city the size of Berlin releases the same amount of microfibres equivalent to 500,000 plastic bags – every single day!

Protect marine animals and the ocean ecosystem by only washing your clothes when they start to smell.

It'll also make them last longer, as less time in the washing machine means less damage and fading.

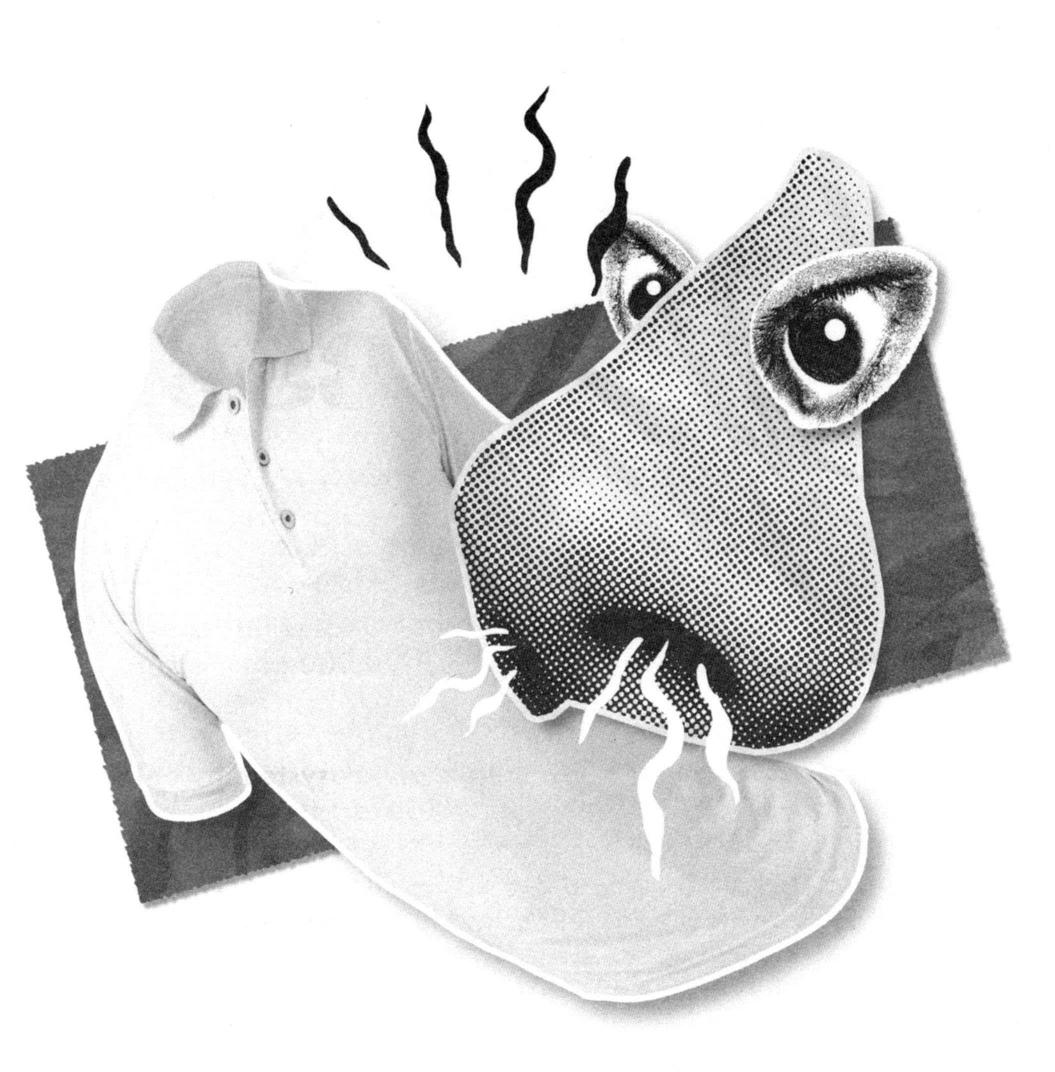

Keep it cool

Want to know another dirty truth behind keeping our clothes clean?

About 90% of the energy our washing machines use goes towards heating the water. If a household switched to cold water washing for a year, they would save enough energy to charge an iPhone 30,861 times.

Plus, you don't need to turn up the heat to get things clean, and washing your clothes too often and too hot can actually be harmful to the garment itself, shortening its life and leading you to have to purchase new items.

Washing in cold water saves energy, saves your clothes and could help save the planet.
So turn down the heat and keep it cool!

Make do and mend

Un-fun fact: we're buying 60% more clothing than we did a decade ago, but we're now keeping each item for just half the length of time.

We know fashion moves fast, but if you can bear to wear that cute hoodie or cropped tee for two years instead of one, you'll actually reduce its carbon footprint by 24%.

And when those clothes get a bit worn out or tired-looking, why not put your sewing skills to the test to fix and refresh them? It can be a lot of fun and you'll end up with some great custom pieces that are guaranteed to turn heads – and change the world for the better!

Start a swap shop

Now you know what to buy, and when and how to wash it, what about the clothes you don't want any more?

That jumper collecting dust in the bottom of your drawer? Don't chuck it – your pal might want it!

Worldwide, 92 million tonnes of unwanted clothes get sent straight to the tip every year.

So, why not start a swap shop? Trade your unwanted clothes with your friends and get new outfits to refresh your wardrobe at the same time.

Give hemp a chance

Meet the fabric of the future – hemp.

It uses less than a third of cotton's water requirements, doesn't rely on pesticides and even replenishes soil quality!

If you've exhausted the second-hand shops and want a new jumper or trousers – try hemp. This sustainable super-fabric is far less prone to shrinking than cotton, so the risk of heartbreak caused by shrinkage of your favourite top from a (rare) laundry mishap from mum is far reduced.

Start a fashion revolution

Sometimes, sitting at home on our phones scrolling through negative climate news can make us feel powerless. But there are small ways you can make a big difference.

Try volunteering at your local charity shop or thrift store, serving customers and sorting through second-hand items.

Not only will you be promoting a sustainable approach to fashion, you'll also be raising money for charity and get first dibs on your fave vintage clothing.

Make it make(up) sense

Now you've got your slick, sustainable, second-hand outfit sorted, it's time to add some sparkle to your (already beautiful) face with some makeup!

The environmental impact of the global cosmetics industry is less-than-beautiful, sadly, due to its love of toxic chemicals, single-use packaging, and the practice of animal testing, which still happens in 80% of the world.

So always check the label, choose cruelty-free and eco-friendly products, and try to buy less by making your makeup multipurpose: an eyeshadow palette can easily double up as a highlighter, and tinted lip balms also work great as blusher.

Chapter 3

Food

Make meat a treat

Rearing farm animals and turning them into burgers, nuggets and hotdogs generates more emissions than all the planes, cars, trains and boats combined. It also drives deforestation, pollution and biodiversity loss.

Reducing your meat intake is one of the most effective ways of limiting your impact on the environment. But going fully veggie or vegan is a big ask, so a great way to start is only eating meat on the weekends.

Your favourite weekday meals can easily be made meat-free with some quick substitutions. Why not swap the mince in your bolognese for lentils, switch out the chicken in your stir fry with plant-based pieces, or opt for vegan bangers and mash?

Ditch dairy

Mass producing cheese and milk has created a major methane problem.

It's 30 times worse than CO2 when it comes to warming the planet and comes mainly from cow burps and cow poo.

13 of the largest dairy operations in the world produce as many emissions as the entire UK. That's a lot of burps!

Switch to plant-based alternatives like oat milk for a cooler, belch-free earth.

Finish your leftovers

We're treating the world like an all-you-can-eat buffet.

When food ends up in landfill, it releases nasty gases like methane; the same planet-warming gas that comes from those cow burps.

Those cauliflower leaves? Roast them.
Those scraps of veg at the back of your fridge? Make a soup.
That leftover chilli con carne you're a bit bored of?
Have it with a jacket potato instead of rice.

Swap beef for beans

You know what's beefy? The amount of water it takes to make a burger. Skipping a beefburger for lunch could save 6,000 litres of water – the same as not showering for 3 months.

Take a leaf out of Jack and the Beanstalk and swap your cow for magic beans!

Packed full of nutrients, beans are your best friend when it comes to eco-friendly protein. They require just a tenth of the water needed to produce beef, one twentieth of the land and release 50 times less CO2.

Always read the label

You are what you eat – or so the saying goes – but do you ***really*** know what's in your food?

A good rule of thumb is, the longer the ingredient list, the more processed the product (and the more nasties potentially hidden inside).

Look out for controversial ingredients like palm oil (roughly 50% of packaged foods contain this leading driver of deforestation), as well as chemicals that are bad for you like sodium nitrate (which has been linked to cancer), refined vegetable oils (which can cause heart disease) and artificial colours, flavours and sweeteners. Yuck!

Are you too clingy?

Cling film... cellophane... whatever you call it – covering your food with single-use plastic film is a seriously eco-unfriendly way to store your leftover curry.

In the UK we use over 1.2 billion metres of it every year. That's 745,000 miles – enough to go around the circumference of the world 30 times over! Studies also show that chemicals found in these plastics can leach into your food and may be bad for your health.

Thankfully, there are some great eco-alternatives out there to stop your sandwiches going stale, including wrappers made from vegan-friendly, wax-coated fabric in a variety of cute colours and designs.

Pick your own

With plastic-wrapped packs of fruit and veg from all across the world filling our supermarket shelves, it might not surprise you to learn that carbon emissions from transporting food make up 3 billion tonnes of CO2-equivalent per year, which is about 6% of the global total.

A great way to cut down on food miles is to get involved with a local pick-your-own scheme. Many farms and fruit growers now invite members of the public to come and help with the harvest for a small fee, allowing you to take home what you pick, packaging-free.

And trust us, it tastes even better when you've worked for it!

Can you peel it?

Did you know, a stomach-churning 2.5 billion tonnes of food is wasted each year worldwide? That's 40% of all food produced for human consumption, and more than 5.8 trillion meals being needlessly wasted every single year.

It's more important than ever, then, that we try to make the most of the food we buy, and there are some really delicious ways to do this, like this simple fruit cordial using just leftover orange peel.

Soak 2 orange peels in boiling water to remove any wax, then rinse thoroughly before blitzing in a food processor with 500ml water. Put the mixture into a pan with 2 cups of sugar and bring to the boil for 10 minutes. Let this cool for 4–5 hours, then add the juice of 2 lemons and strain through cheesecloth. Serve chilled for a delicious, nutritious, planet-saving treat!

Grow your own

A wise person once said, 'Mighty oaks from little acorns grow...' and it's a good phrase to remember when it comes to our own efforts to save the planet.

And on the subject of gardening, a small (but also big) way of making our food system more sustainable is to try growing your own fruit and veg at home!

So ask your family if you can plant a small fruit or vegetable patch in your garden, if you have one, or even in pots on a windowsill or balcony. It's great fun to muck in together and food tastes even more delicious when you've grown it yourself.

Take-away the polystyrene

Nothing beats chip shop chips. Or a takeaway veggie burger. But unfortunately, they're often packaged in polystyrene. Ask your local kebaby to put it in a cardboard box or wrap it in paper instead.

A single tray of polystyrene takes over 500 years to decompose. So that's a lot of waste that won't be going anywhere, anytime soon!

Oh! And say no to disposable cutlery; it always ends up in the bin anyway! And if you want to be a real friend to the planet, you could bring your own.

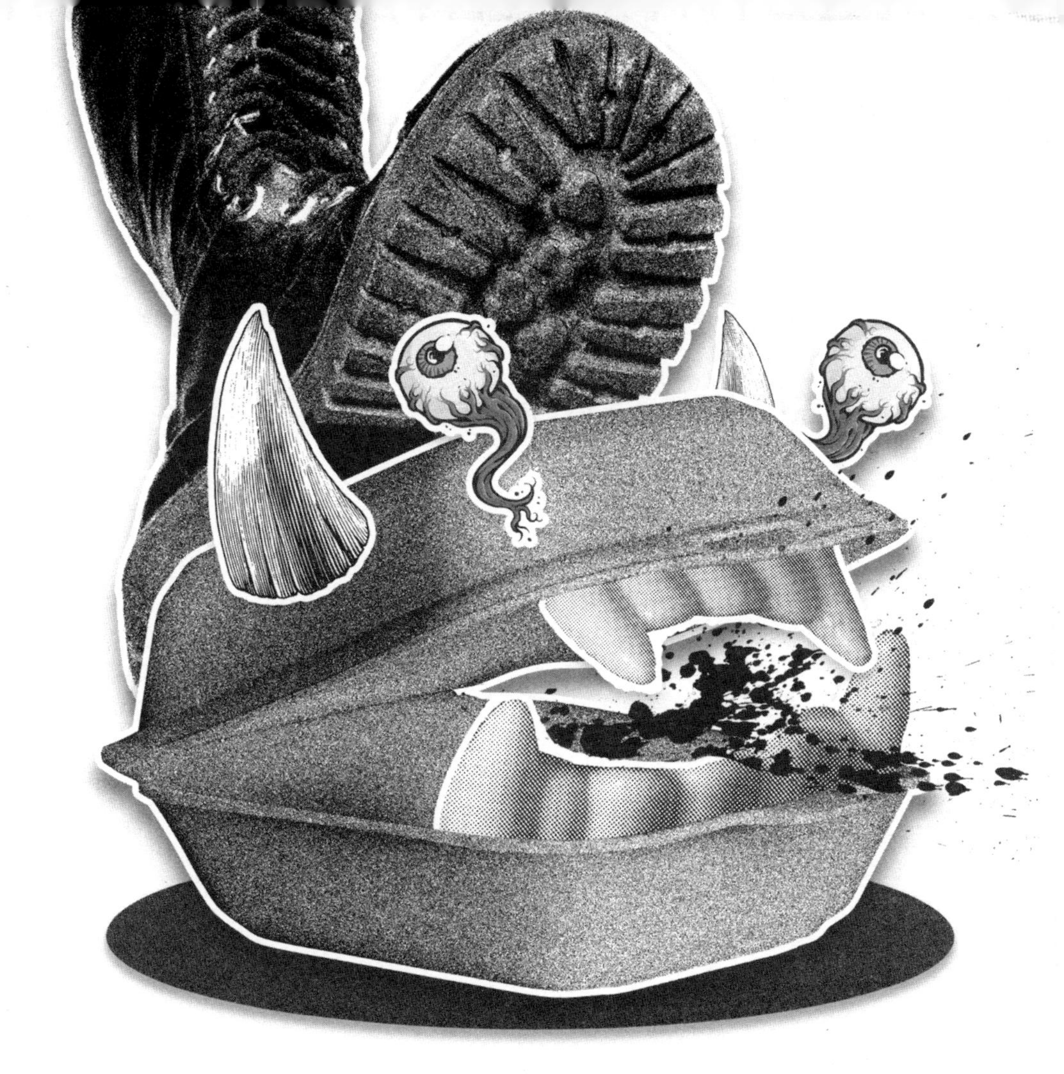

Oompa Loompa Doompadee-DON'T

Don't get us wrong, we love chocolate just as much as the next person, BUT...

In 2020, the Ivory Coast lost 116,000 acres of forest because of cocoa farming. The chocolate industry is also notorious for child slavery on cocoa farms. These farms provide chocolate for the largest chocolate distributors in the world... so be careful what you're buying.

Make sure to buy from a brand known for being sustainable and 100% slave-free for a delicious, ethical and guilt-free snack (or snacks, we don't judge!).

Eat like a fish

...Not literally; you'd probably get some weird looks.

We're talking about **SEAWEED**.
Seaweed is not only super nutritional and good for us to eat, but it also grows quickly, provides a habitat for fish (and otters) and absorbs carbon.

Buying seaweed or products that use it will help to support seaweed farms.

The more seaweed we grow the better our oceans will feel!

Chapter 4

Home

Pee on your leg

Yep, you read that right!
Okay, we get it, this one might seem a bit weird, but hear us out...

Flushing the toilet uses, on average, 7–8 litres of water per go, and showering uses a whopping 12 litres of water per minute. With the average shower being around 10 minutes long, that's up to 150 litres of water.

We've crunched the numbers, and the best solution is (drum roll, please)... combining the two and having a wee in the shower! Once you get used to it, you might even enjoy the feeling of freedom it gives you, and you'll use a lot less toilet paper too. (Probably best to save the Number Twos for the loo, though!)

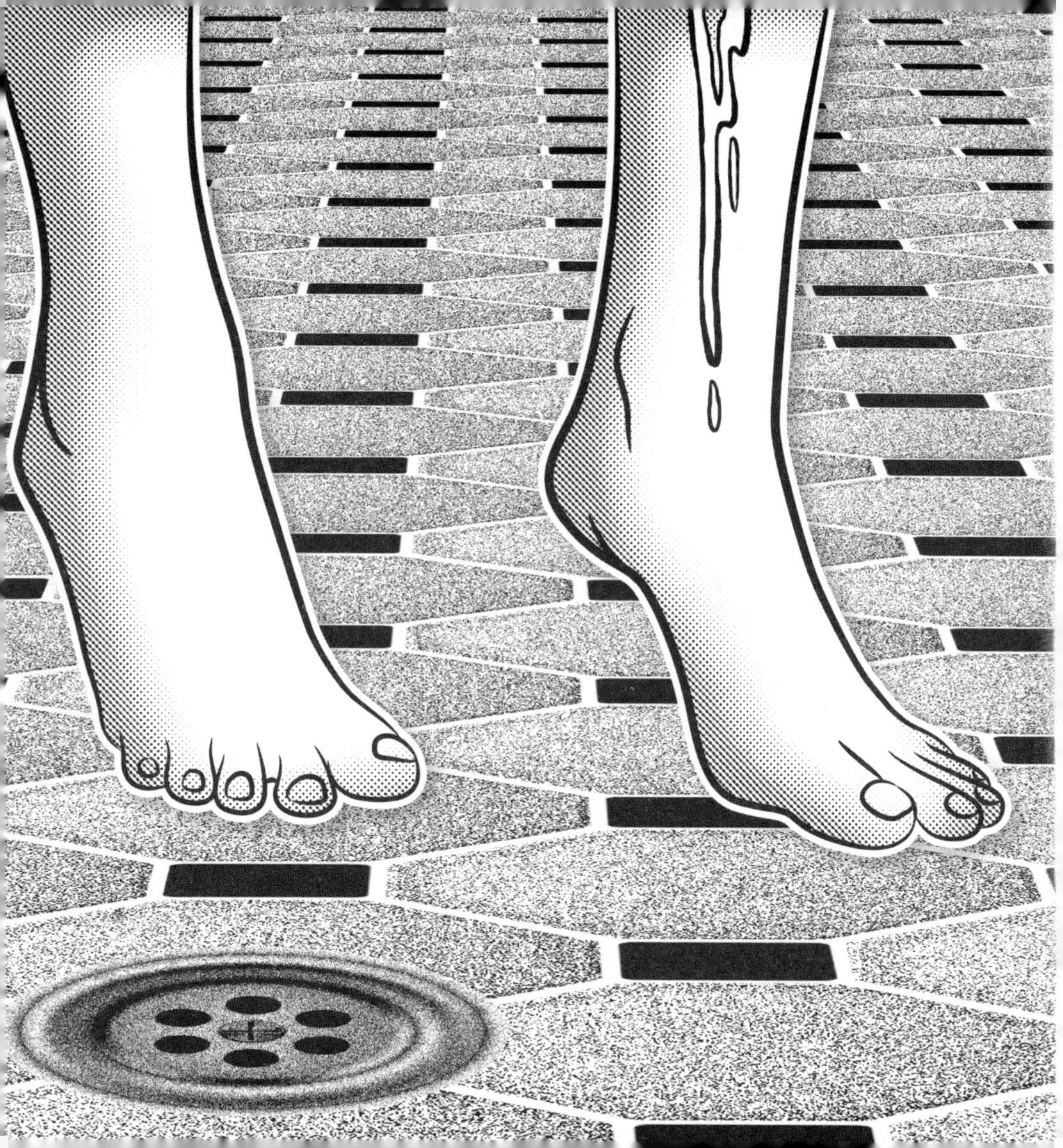

Clean your rubbish

You might not look at baked beans, yoghurt pots or juice cartons as treasure – but if recycled properly, the packaging they're in is priceless.

The only problem is rogue beans can contaminate the rest of the recycling. Which is partly why only 9% of our rubbish gets recycled worldwide.

Rinse and dry your rubbish to give it the best chance of being turned into more delicious snack casings. Because the world needs more beans (not beef)!

Switch off

Did you know that even when you turn a TV off it's still using power?

Standby power accounts for around 9% of our home electricity use. To bring that down to 0%, turn your electronics off by the switch when you're done.

Your appliances aren't the only things you should switch off – switch yourselves off too! Take a break from your TV and reconnect with nature. The planet needs you refreshed!

Bamboo your poo

Bamboo isn't just great for pandas.
It's a fast growing, versatile wonder plant.

So it seems a bit crazy that we cut down 1.9m trees daily just to use for loo roll. That's enough trees to fill Wembley Stadium 50 times over!

Give the lungs of our planet a breather and swap your toilet paper for bamboo rolls.

Turn the taps off

Do you know how much water we waste just by leaving the tap running while we brush our teeth? An average of 4 gallons each time!

So how can we reduce this terrifying stat? Simple. Turn off those taps! If you aren't using the water in between brushes then there is no need to have it on. Save that precious water when brushing those pearly whites.

Oh, and make sure the taps are completely turned off when you're done!

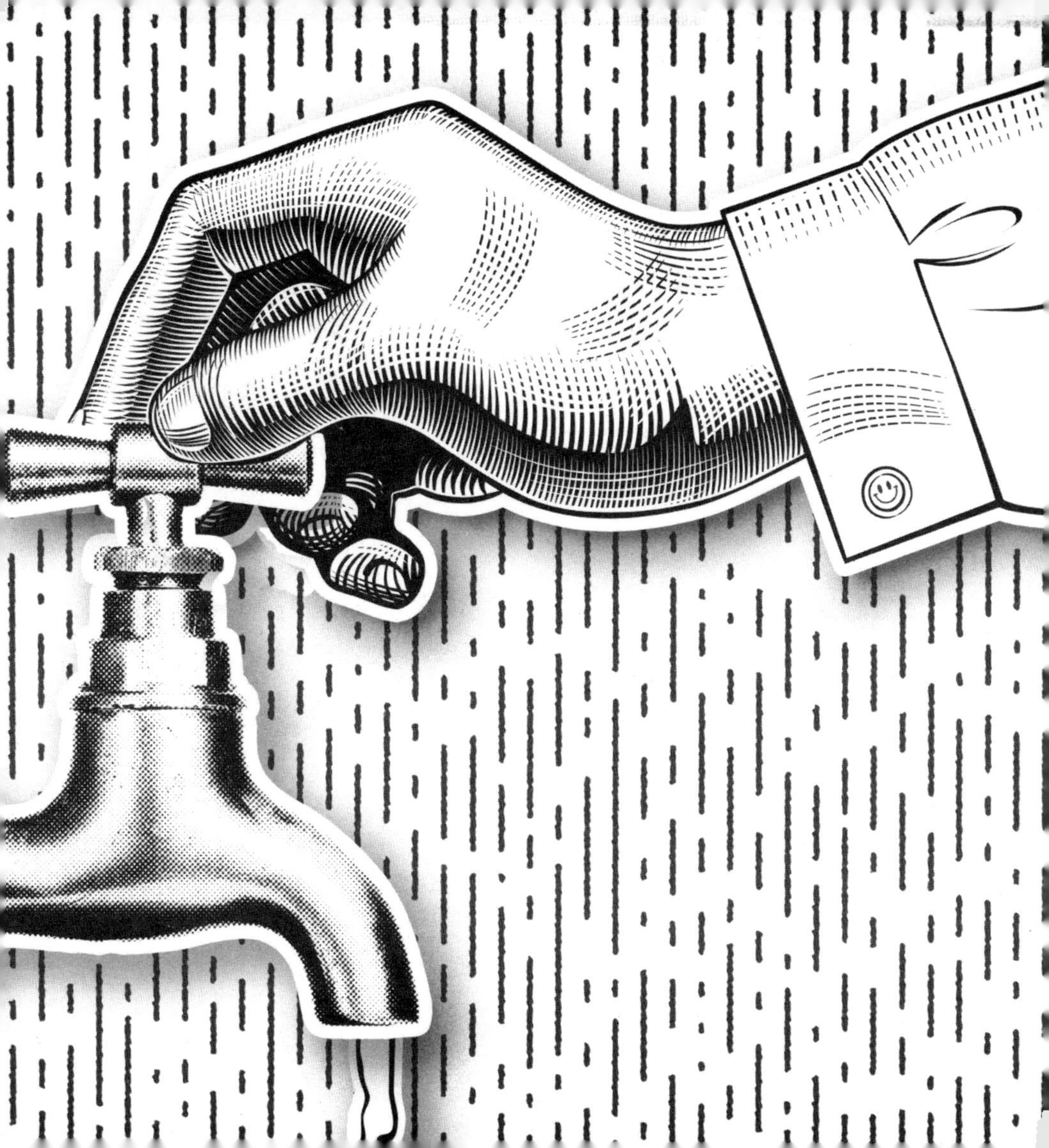

Wax lyrical

Candles are a cute and affordable way to brighten up your bedroom. Sadly, they can also be bad for the environment and bad for your health.

Most commercially manufactured candles are made from paraffin wax, which is derived from petroleum and releases acetone, benzene and toluene when burnt – chemicals that have been linked to cancer.

But there's no need to sit there in the dark! There are some great alternatives out there, including rapeseed wax, which is made from plants and is renewable, non-toxic and burns clean. Even better, why not get crafty and make your own eco-friendly vegetable-wax candles using a kit at home?

Put a hoodie on

'If you want to be a goodie, put on that hoodie'
– Your Nan

But seriously, she's onto something. Turning up your thermostat or switching on an electric blanket releases emissions. So throw on a jumper first to warm yourself up. You could even ask your Nan to teach you how to knit one for extra eco points.

Add some slippers and make yourself a hot chocolate to complete the look.

Turn it down!

Now you're all snug in your Nan's knitted jumper – maybe it's time to actually turn the heating down a bit?

Lowering your thermostat by just 1 degree will save your family money on the heating bill. It also prevents around 300kg of CO_2 being released per year, the same emissions as driving 745 miles in a petrol-powered car!

Donate your old toys

Times were much simpler when you were a kid playing with toys on your bedroom floor.

Now that they're collecting dust in a box, why not donate them to a younger relative or a charity shop?

3 billion new toys are brought every year in the US alone, most of which are made from virgin plastic, so for the sake of the planet, we should start sharing our favourite toys more.

Mellow yellow

Who doesn't love a bit of toilet humour? Especially the old rhyme that goes, 'If it's yellow, let it mellow, if it's brown, flush it down' – in fact, we reckon everyone should make this their mantra!

Due to climate change, by 2050 the amount of water available to us could be reduced by 10–15%, so wasting it is a big no-no. If you can't wee in the shower, leave it in your loo.

It's actually perfectly clean (and much better for the environment) to leave your wee in the toilet and only flush when you've done a Number Two.

Worship the upcycle

We've got a serious waste problem. 99% of the things we buy end up in the bin within 6 months. That means globally, we throw away 2.2 billion tonnes of rubbish every year.

To stop our landfills, oceans and streets filling up with waste we need to repurpose it.

Like using old toilet rolls to stop your cables from getting tangled, or adding a bit of paint to an empty can of beans so it becomes a flower pot.

It saves money, reduces waste and is one more step towards saving the planet!

Put a lid on it

Next time you're boiling water for some instant noodles, it's worth remembering that it takes twice as long to heat with the lid off than with the lid on.

On average, electricity and gas use creates about a quarter of all carbon emissions from our homes.

So keep a lid on your pan and you'll reduce the amount of heat and time required to cook your food (those extra seconds count when you're hungry!), cutting down on energy waste and your household bills too.

Don't sweat it

With global temperatures rising every year and heatwaves becoming more and more common, the temptation is to set the air-con controls to 'arctic' and enjoy some chill time.

However, air conditioning actually makes the climate problem worse, consuming 10% of all global electricity during peak times and leaking nasty planet-warming gases called HFCs into the atmosphere.

But don't sweat! You can make your own DIY eco-friendly air-con by putting a tray of ice cubes in front of a normal fan to send a fresh and frosty breeze throughout the room.

Stop pouring water away

What's one thing humans simply can't get enough of? WATER. So why are we pouring so much of it away?

Every day we waste over 3 billion litres of water. That's enough water to fill 1,200 Olympic-sized pools! Feeling thirsty yet?

Leave out buckets to collect rainwater and use that to water your plants. Fun fact: tap water has a level of chlorine in it that can prevent plant growth. So by using rainwater, we're helping our plants out and using less water!

Get some worms

If you have access to any kind of outdoor space – big or small – making a compost bin is a great way of turning leftover food into a nutritious, organic meal for your plants – and a home for your friendly neighbourhood worms!

When your food waste gets sent to landfill, it decomposes anaerobically (without oxygen) and creates powerful greenhouse gases like methane that contribute to climate change. But composting at home allows organic waste to break down naturally without harmful gases, and it makes great fertiliser for your garden.

It's super easy to get started, and research shows that composting at home for one year can save the equivalent of all the CO2 your kettle produces annually.

Chapter 5

Online and Tech

Search green

Want to save the planet while you search the web?
We've got you covered!

Ecosia (ecosia.org) is a 'climate active' search engine that uses the ad revenue from your searches to plant trees in more than 35 countries around the world.

To date, they have worked in tandem with local communities to plant over 200 million trees, including over 500 different native species to protect the natural biodiversity of different areas.

Ecosia puts all their profits towards climate action, with at least 80% financing tree-planting projects around the world, and their solar panels producing enough energy to power your searches twice over (so more renewables and fewer fossil fuels – hooray!).

Be smart with your phone

Ring, ring, reality calling! Let's admit it, we're all smartphone addicts, but our hunger to have the latest handset is a habit we really need to kick.

Whilst most phones are built to last at least four years, 51% of us admit we upgrade every two. That's pretty wasteful when you realise that manufacturing each device clocks up an estimated 67kg in carbon emissions, not to mention the precious metals that are mined to make them.

So take care of your phone and make it last – and who knows, if you keep it for long enough it might just become vintage chic!

Unfollow the leader

Social media is a great way of staying up-to-date with all the latest eco news and for building movements that can change our world for the better, but as we all know, it also has a darker side.

We're constantly being targeted with misinformation (climate change deniers, anyone?), as well as ads for things we really don't need (I know, I know, that Pikachu rug really tied the room together).

We're only human so sometimes it's hard to resist, but one of the best things you can do is simply unfollow social media accounts that are spreading fake news or fast fashion accounts that tempt you to buy things you can do without. It really is that easy!

-1
21,098
56,793
100,897

Bank green

When it's time to set up a bank account, think carefully about which bank you choose, because you'll also be choosing where your money gets invested.

Since 2015, the 35 biggest banks in the world have jointly contributed $2.7 trillion to fossil fuel companies, in spite of pledges to achieve net zero emissions in the near future. Now that's what we'd call dirty money!

So do some research (https://bank.green is handy), and make sure you choose a bank that is committed to investing in green, ethical projects, and avoid any that invest heavily in polluting industries like fossil fuels, mining, tobacco firms or those that test on animals. Money talks!

Show some love for old tech

We get it, brand new shiny things can be very enticing.

That tablet you've seen on all the adverts – you want it. So why not buy refurbished?

You'll be fighting e-waste (85% of which ends up in landfill), and you'll be getting top tech for a fraction of the price. Great for the planet and your pocket!

Don't just like

Climate action often starts online, and social media can be a great resource for learning about our environment and how you can do your bit to help.

So when you see a climate crisis solution or a shocking stat on social media, don't just like and move on – make sure you share it with your friends and followers to raise awareness and inspire others.

Movements are made by people like you, and it's important to use your platform – however big or small – in the fight to save the planet.

The truth is out there

Although social media can be an incredibly powerful tool for good – especially when it comes to raising awareness about the environment – it's important to remember that some people also use it to spread false information and to skew the truth.

Unbelievably, there are people and organisations out there who deny that climate change is real, and want to stop our governments from taking it seriously and taking action.

So make sure you always check your facts and do your research when it comes to the things you read online – even if they seem to back up what you already believe.

Clear your inbox

What's your unread email count?
Are you an inbox zero kinda person, or do you live for the chaos of keeping that number in quadruple digits?

Either way, it's important to remember that all those emails from Deliveroo, Asos and Playstation have a carbon footprint too, due to the energy it takes to store them in huge data centres. Over the course of a year, each of the world's 3.9 billion email users can average up to 40kg CO_2e, which is the equivalent of driving up to 128 miles in a petrol car.

So, how can you reduce your digital carbon footprint?
Start by sending fewer emails and send links to documents rather than attaching them. Unsubscribe from mailouts you don't read (yes, even that constant stream of Sale Now On emails you get from a fashion brand you bought a solitary sock from five years ago). And above all, delete, delete, delete – clean inbox, clean mind!

Donate your old mobile

Believe it or not, mobile phones can outlive their 24-month contracts. Shocking, we know!

Keep your phone as long as you can, and when it's finally time to upgrade, don't just chuck it away. In the US alone, 151 million mobiles are incinerated or sent to landfill each year.

When your phone has run its course, try donating it to charity. You'll be helping out someone in need and be preventing e-waste, which we produce enough of every year to outweigh the Great Wall of China.

Shout about it!

When you've made a positive change in your life or done something to help the planet – however big or small – remember to tell your friends, post on social media about it and spread the word!

Did you wash out your recycling?
Did you remember to bring your tote bag to the shops?
Did you donate your loose change to an eco-charity?

Your actions will inspire others to take action, and that's how we change the world. So don't be modest – make a noise and shout about it!

Get ticked off

Congratulations, you've made it to
halfway through the book!

You're probably thinking how on Earth am I going to
remember to do these planet saving tips in my day-to-day?

The answer is make a to-do list. Set yourself some achievable
actions each week so you don't feel overwhelmed by too
much change e.g.

1. One meat-free lunch this week ✔
2. Ask a friend if they're worried about the climate crisis ✔
3. Share a climate solution on social media ✔

Chapter 6

Shopping

Stop buying stuff

One of the most sustainable things you can do is nothing... well, at least when it comes to buying stuff!

Overconsumption is depleting our natural resources and destroying nature. Only 1% of products are in use six months from their purchase.

So make the most of what you already own to reduce emissions. And it saves you money too – result!

Be thirsty for change

Thinking about single-use plastic water bottles makes us thirsty for change!

As well as the huge amount of emissions and waste they cause, there's growing evidence to show that plastic water bottles contain chemicals that may be harmful to human beings. Studies have also found traces of pharmaceuticals, microplastics, bacteria and heavy metals in several popular bottled water brands.

A lovely glass of filtered tap water is a far more refreshing, sustainable and healthy alternative. Stay hydrated, kids!

Boycott bad brands

Did you know 100 companies are responsible for 71% of global emissions? Beyond that, some of the most well-known brands in the world are also the biggest plastic polluters, chemical users and drain on natural resources.

When you hear about the latest brand destroying forests, mistreating workers and polluting the ocean – take a stand.

Don't give them any more of your money and switch to an alternative.

Ban the balloon

Sorry to burst your bubble, but balloons are bad news and there's not much to celebrate when it comes to their impact on our environment.

Even biodegradable latex balloons can take up to four years to break down, and in the meantime they end up polluting nature and harming wildlife that often mistake them for food.

There are plenty of alternative ways to mark a special occasion – what about planting a tree or using your new upcycling skills to make bunting from leftover wrapping paper?

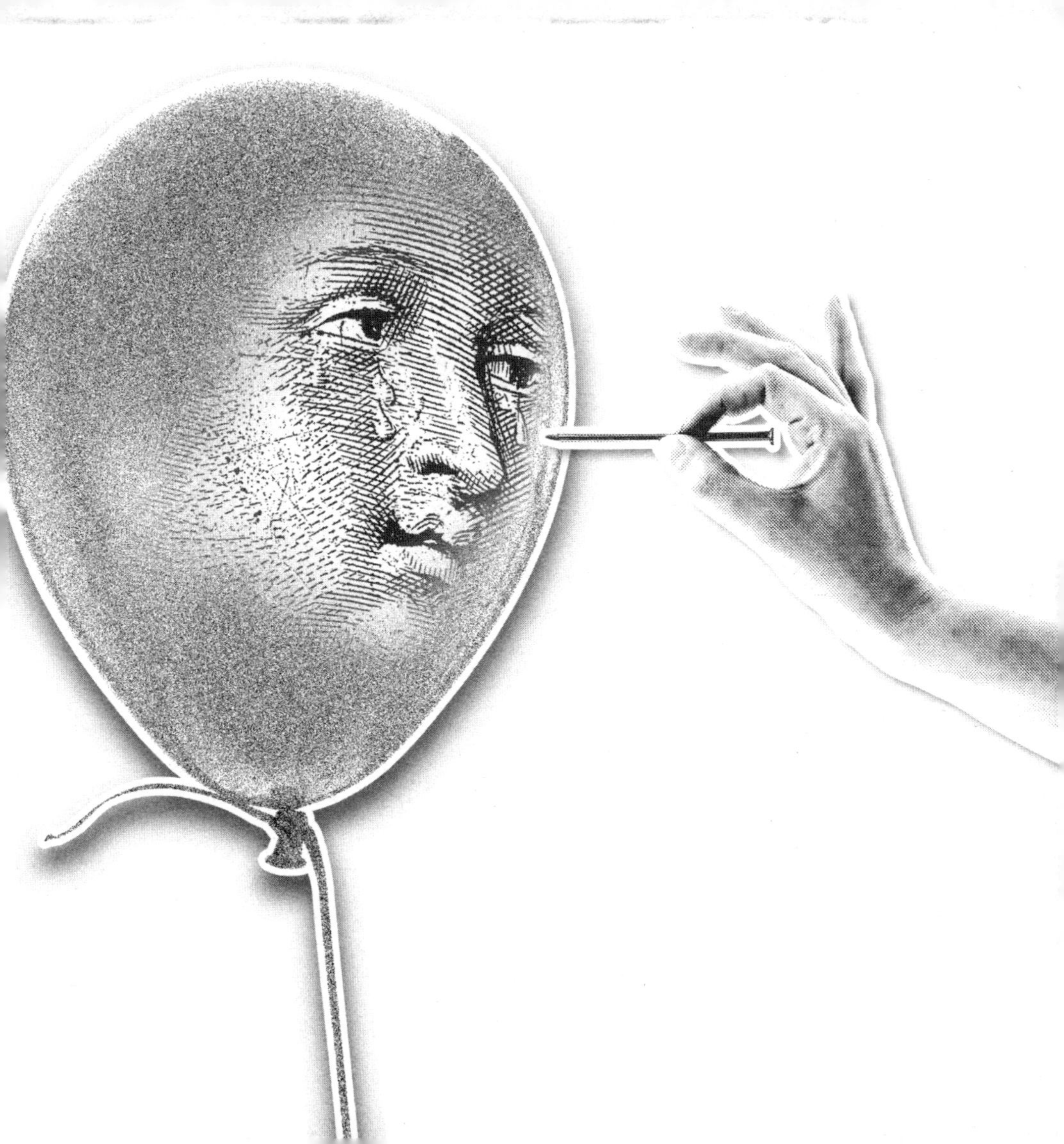

Bring back green Santa!

Did you know that Santa Claus used to wear green until 1931, when a Coca-Cola advert popularised the red suit we all know and love? Back then we gave fewer Christmas presents too.

Lovely as it is, the festive season can also be a time of excessive consumption and extreme waste, with 81 million unwanted presents given in the UK each year – not a very merry thought!

So take a leaf out of Green Santa's book and get creative with your gifting by choosing lower-impact items for friends and family. How about a membership to a local gallery or museum, a donation to an eco-charity, or perhaps you could even pay to plant a tree in their name?

Bulk up and refill

141 million tonnes. That's the amount of plastic packaging produced worldwide every year – the same weight as 21 Great Pyramids of Giza!

From pasta, rice and spices, to shampoo, shower gel and toothpaste, there are certain staples in life that we need to keep topped up. But most of these items come individually packaged, which contributes to that humungous pyramid of plastic.

To make your shopping more sustainable, consider buying in bulk. Larger quantities last for longer, so less waste goes back into the world. And if you've got a zero-waste refill shop nearby, you can even take your own reusable containers along and top up your staples there – it's a packaging-free paradise!

Find natural alternatives

Many big brands use chemicals that affect your hormones and leak into our waterways hurting marine animals.

But have no fear, there are SO many alternatives out there that are natural and healthy. There's even some home remedies you could try! Squeezing lemon juice into a spray bottle makes for a great dry shampoo, and even mixing apple cider vinegar with water gives your hair strength and volume without the need for chemicals!

It's so easy to buy the first product you see on the shelf. But do you really want to be putting those chemicals into your body? We'll take staying natural over hurting marine animals any day!

Ditch aerosols

Binning aerosols might just be the best thing you do this year!

When we breathe in those tiny particles, they can do nasty things to our lungs whilst also contributing to air pollution.

Luckily for you, there's loads of alternatives that won't contribute to global warming. Take natural roll-on deodorants that are in paper or metal containers, you won't be spraying any of that toxic spray into the air and you'll still smell just as fresh.

And why not encourage your friends to do the same? Remember: say it, dont spray it!

Shop local

We've all been there, casually browsing online, adding things to our basket, changing our minds... then buying them all anyway because it's so easy and we can (and probably will) just return them.

However, home deliveries and returns from e-commerce clock up huge carbon emissions, with 3 billion trees cut down each year to produce packaging, meaning this isn't really a sustainable way to shop.

Instead, why not check out your neighbourhood stores and shop local? This helps to reduce the carbon footprint of the products you buy, and you'll be helping to support independent businesses in your community too.

You CAN do it

Diet or regular, fizzy or still, when it comes to quenching your thirst we've got some refreshing news for you: aluminium cans are way better for the environment than plastic or even glass bottles, and are infinitely recyclable!

Even better, recycling one of these little beauties saves 95% of the energy used to make a new one and no extra material needs to be mined or transported.

We reckon drinks are more delicious out of a frosty-cold can too... but maybe that's just the sweet taste of saving the world!

Get barred

Getting barred isn't something we'd usually recommend, unless it's swapping your shower gels and shampoos for soap bars and shampoo bars of course.

Studies have found that around 238 million Americans use body wash instead of bar soap – that's about 1.4 billion disposable bottles a year!

Not only will you be reducing plastic waste by opting for soap bars, but they're less likely to be full of the nasty chemicals that many shower gels contain, meaning that they're better for your body too!

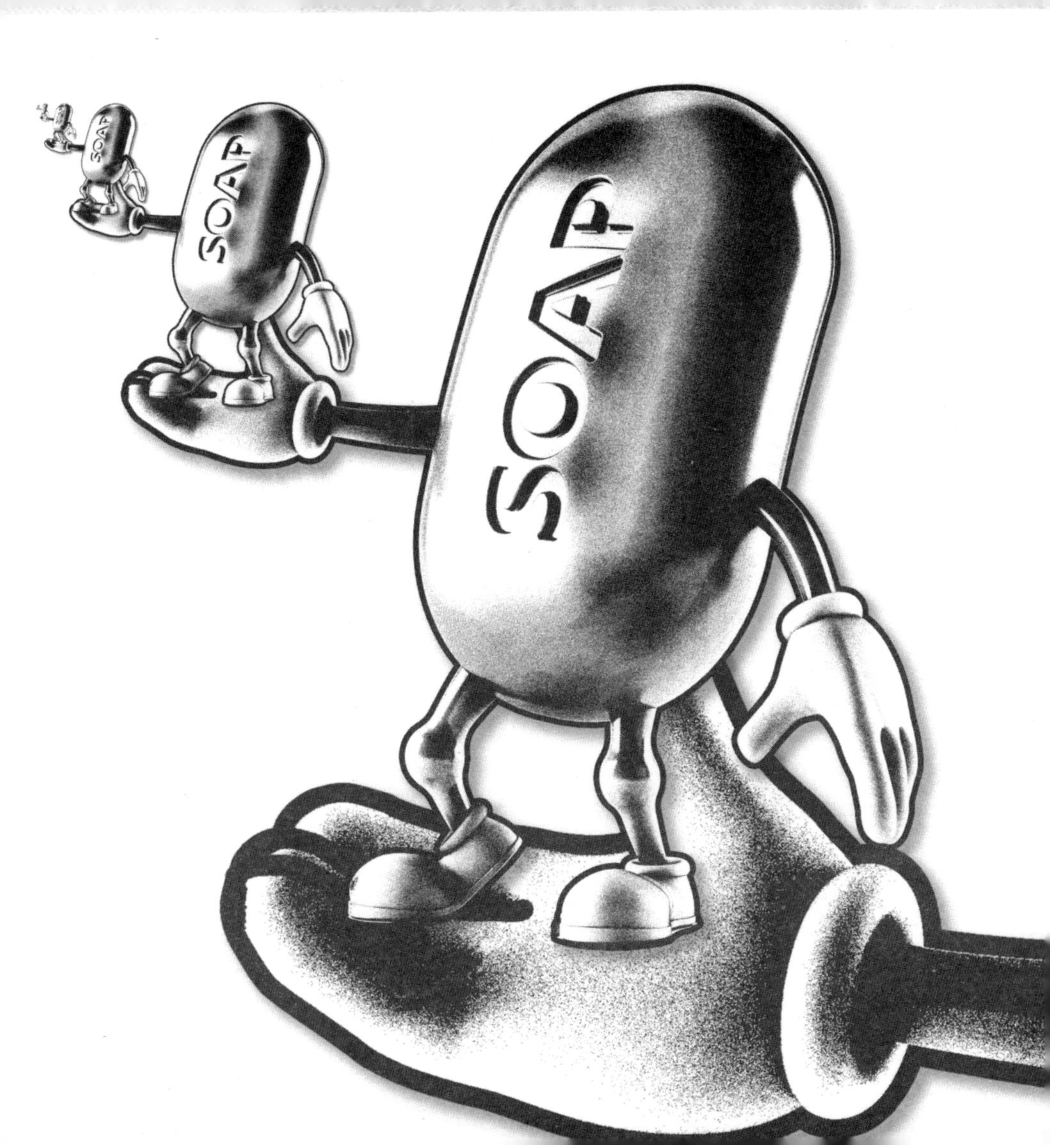
SOAP
SOAP
SOAP

Choose to reuse

Plastic is a bit like Monday morning: it's annoying and it seems inevitable. But there is another way. Start your plastic-free journey by ditching single-use plastic for reusable alternatives.

Try using a bamboo lunchbox or beeswax wraps instead of clingfilm when storing food, and switch to a reusable coffee cup or metal bottle for your drinks.

These must-have accessories come in all shapes, sizes and colours, so you'll definitely find something to suit your style!

Turn your period green

Girls, we get it – when it's that time of the month, the last thing you want to hear is that your period products are unsustainable, but sadly, that's the headline.

On average, girls and women throw away up to 200kg of sanitary products in their lifetime, and with most containing up to 90% plastic (not to mention a cocktail of nasty chemicals, which can be bad for your body too) it's enough to give you permanent PMT.

For an eco-friendly period, why not try one of the growing number of reusable or biodegradable alternatives, like menstrual cups, washable pants with a built-in lining, or compostable pads and tampons? And don't forget to treat yourself to a tub of (dairy free) ice cream while you're at it!

Be totes amazing

Brace yourself for some depressing stats about plastic bags... Globally, we use 5 trillion plastic bags per year. It takes 1,000 years for a plastic bag to completely break down, but on average, each one is used for just 12 minutes!

We could go on, but hopefully this is enough to help you remember to take your reusable carrier bags or fabric cotton or hemp totes out with you next time you go shopping.

It's important to remember that tote bags do have an impact on the environment too, but if you use them enough (172 times to be precise) then they can make a real difference.

That's a wrap

Not to be a Grinch, but while we all love gifts wrapped in shiny paper and bows, one of the biggest festive waste products is wrapping paper.

More than 11,000 tonnes is used each year in the UK alone – the equivalent of over 60,000 trees – and most is single-use only, often containing non-recyclable additives like metallic foil, glitter and plastics.

Always make sure you look for fully recyclable paper when buying, try to reuse last year's paper when you can, or even get creative and use colourful fabric to wrap your gifts (tied with string to avoid using sticky tape).

Chapter 7

Nature

Hug a tree

'Tree hugger' has got to be one of the lamest insults. We're proud to hug trees, and you should be too.

Studies have shown that regularly being amongst nature works wonders for our mental and physical health. It reduces stress, helps protect against depression and improves our mood. It's also been shown to create a closer connection to how our climate is changing.

So how can you get more connected to nature? Try hiking, bird watching, get into gardening – the list goes on!

Make friends with spiders

Cute or scary, there's one thing spiders are good at: balancing nature's ecosystem.

They rid your home of pests – 800m tonnes of them a year to be precise – so that means fewer disease-carrying mosquitos, and moths that feast on your clothes.

So next time you spot a spider indoors, don't stamp on it or throw it out the window if you're brave enough – let them get to work creating a web that'll keep your house pest-free.

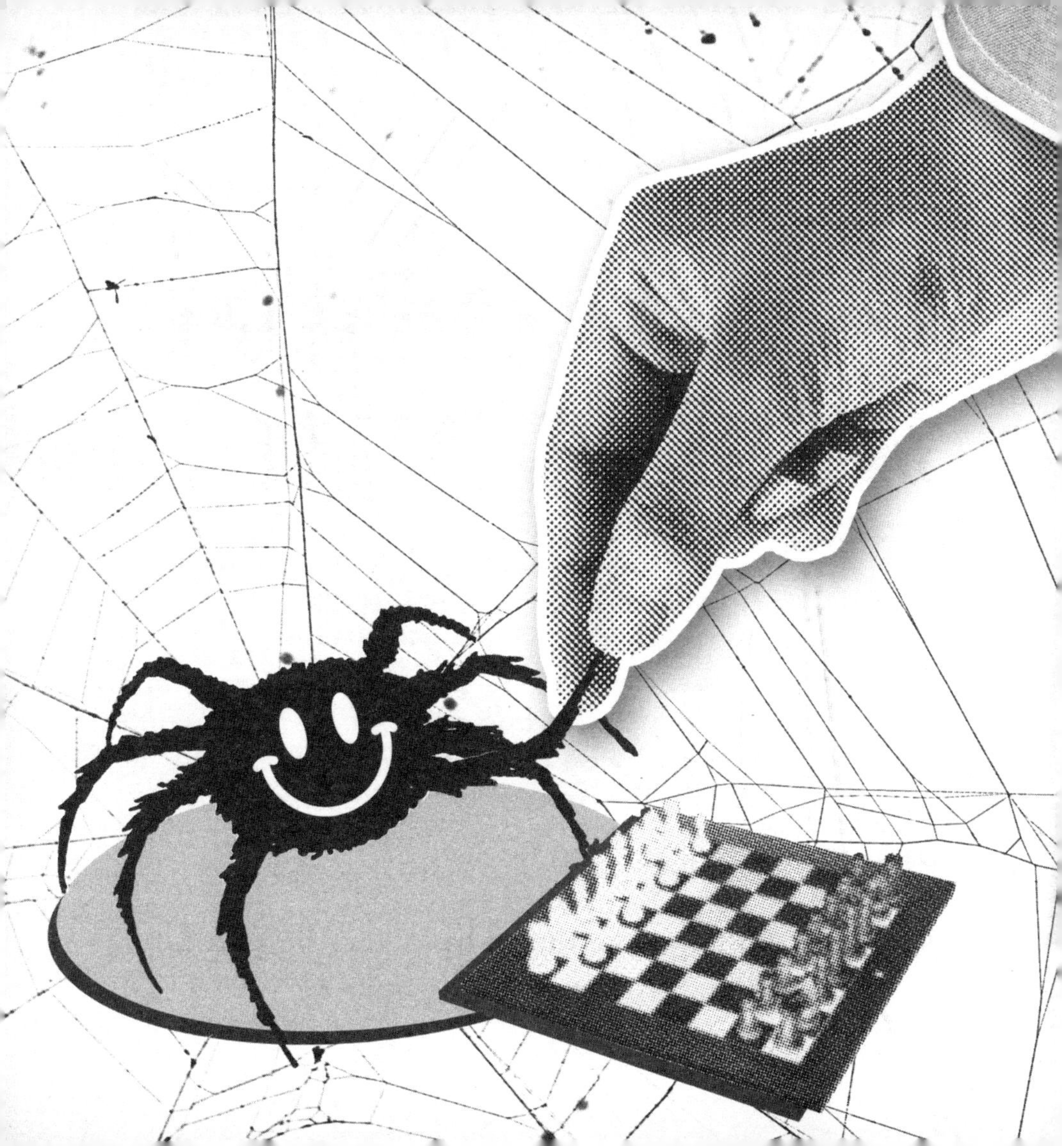

Bee nice!

Bees don't just sting and make honey.

Without their pollenating power you'd be saying goodbye to some of your favourite foods and animals. And due to habitat loss and climate change, their extinction is a possibility.

Make your garden as bee-friendly as possible by filling it with pollinating power and the little buzzers won't stay away. Some of their favourites are bluebells, lavender and primrose!

Or if you're feeling really creative, give the bees some luxury accommodation. With some wood, power tools, bamboo and a bit of elbow grease, your very own bee hotel will be open for business.

Think of the turtles

If there's one thing that turtles love – it's jellyfish.

But there's something else in the deep blue that looks just like a tasty jelly: plastic bags. When you chuck yours away it could end up in the ocean and a turtle's stomach.

This goes for any throwaway plastic item – straws, coffee cups, bottles, or polystyrene – marine life mistake them for food so we need to avoid them and switch to reusable items or ones made from natural materials.

Make the world your gym

Gyms are great and all, but why pay to exercise when you could work out in the world for free?

Added to which, there's the carbon cost of getting there if you go by car or public transport, not to mention all the energy gyms use to keep their lights on and their treadmills whirring.

Getting outside to exercise – be it running, walking or playing a sport – is one of the best things you can do for your mind and your body, and using your local parks and green spaces is a great way to make the most of the environment around you.

Last one to the tree's a loser!

Litter pickin' good!

Change can come from the little things we do, because those little things can eventually add up to become big things!

One great way to do your bit each day – whether you're on your way to school or to hang out with friends – is to challenge yourself to pick up three pieces of litter from the street and put them in the bin.

It might not feel like much, but if we all did it, we'd have cleaner streets and a better environment we could all be proud of.

Adopt an animal

If you love animals, then this one's gonna make for scary reading. A recent report has shown that there are currently 30,178 species in imminent danger of extinction. Extinction means gone. Forever.

With so many threats to wildlife – including destruction of habitat, hunting, pollution, climate change and more – it can feel like a hopeless situation. But there is a way you can help!

Charities like WWF make it easy to sponsor or 'adopt' an animal with just a small donation, and your money will go directly towards restoring their habitats and protecting wildlife. This can also make a wonderful – a doubly sustainable – gift for the animal lover in your life!

Feed the birds

Put a bird feeder in your garden with organic nuts and seeds.

Birds play a vital role in our ecosystem. Birds like the hummingbird help to pollinate important plants that are used in human medicines, so let's help them to help us!

You could even start bird-watching and see what birds your feeder is attracting!

Note to self: be careful when telling people you have seen blue tits in your garden.

Drop a bomb

Ew, no! Not that kind of bomb...

You can lend a hand to our bumblebees, honeybees and butterflies using a seed bomb.

Seed bombs are basically small balls (hence bombs!) of tightly packed compost with wildflower seeds inside. Drop a bomb where you please and once you water it, or it rains, it'll germinate, bloom and bring a pop of coloured flowers to your grass.

Top tip: autumn and spring are the best times to plant your seed bombs!

Save soil

There's a climate hero lurking beneath our feet and its name is soil.

One teaspoon of soil contains more microorganisms than there are people on Earth.

Worldwide it has absorbed 3,000 gigatonnes of carbon – four times the amount in the atmosphere. It's essential for food production and biodiversity. But human activity is ruining its ability to save us from CO2 emissions, floods and hungry bellies.

Give soil a break by growing your own fruit and veg, then give back to the soil by adding any scraps to a compost heap to create your own nutrient-rich soil.

Chapter 8

Travel

On ya bike!

Drivers are a bit lazy when you consider that 50% of car journeys are under two miles long. You know what that means? On your bike!

Start by cycling to school, college or work instead of getting a lift. The more you cycle, the more you'll improve the air quality in your local area. You also get the added bonus of avoiding traffic, great for when you're running late, and not having to ride public transport.

Cycling gives you a greater sense of freedom, as you can explore lots of different routes whilst getting some exercise in. Invite friends along too to make it a social activity. In Barcelona they call it bicibús (bike bus).

Carpool karaoke

Although it's always better to walk, cycle or get public transport when you're going somewhere, sometimes getting in a car is unavoidable.

But there's still a way to reduce their impact on the environment. Next time you're being taken somewhere in a car, ask yourself if there's a way you can share your ride with a friend or family member who might be going the same way.

The more people you invite to ride in the car at one time, the fewer cars there'll need to be on the road, meaning reduced emissions and less traffic clogging up the roads.

You'll just have to fight over who gets shotgun!

Walk it off

The best way to get your bearings in a local area and find those hidden gems is to walk around it.

Walking burns calories, keeps your heart healthy and if you're breathing in fresh air it will regulate your serotonin levels, making you feel happier.

You'll leave the planet feeling happier too, reducing road noise, air pollution and microplastics that leak from car tyres.

Trip chain

Even if you catch the bus or train on a regular basis, it's still a good idea to minimise the number of journeys you make.

'Trip chaining' can be a great way to do this. It's all about saving emissions by grouping your errands into one journey instead of going back and forth.

And if you need a bit of extra motivation, why not try bundling in an activity you enjoy – like grabbing a box of chips from your favourite takeaway or a tasty oat milk latte – to balance out your chores.

Start a car wash

If all the cars on your street have 'clean me' written in finger-graffiti on their dirty windscreens, your next planet-saving challenge has arrived!

Automatic car washes use gallons of water, electricity and nasty chemicals, but you can help out the planet – as well as your family and neighbours – by grabbing a bucket and a sponge and starting your own eco-friendly DIY car wash.

There's still an environmental impact to car washing, even at home, so you'll need to make sure you use a biodegradable cleaner specifically designed for cars, and wash on a lawn so that the waste water can be absorbed and neutralised into the soil instead of flowing directly into drains and waterways.

Holiday locally

When it comes to holidays, America may have Universal Studios and Miami, but here in the UK we have Alton Towers and Skegness – we know what we're picking!

A return flight from London to New York emits 1.7 tonnes of CO_2. That's a third of the average Briton's yearly CO_2 emissions.

So embrace the staycation. The UK has beaches, Mr Whippys, countrysides and trains to get you there sustainably – what more could you want?

Sunscreen not SunSCREAM

Nobody wants to end up looking like a lobster on the beach (and we're not only talking to our ghostly pale goths out there), so wearing sunscreen to protect our skin is a must for all of us.

Sadly, most sunscreen contains toxic chemicals like oxybenzone and octinoxate, that once in the ocean, harm the beautiful (and irreplaceable) coral reefs that provide a home to around 25% of all the world's marine life.

So to keep your skin sun-kissed and eco-friendly, make sure you check the label and choose a 'reef-safe' SPF that uses physical UVA and UVB filters like zinc oxide and titanium dioxide, instead of nasty chemical ones.

It's not all or nothing

Living a completely sustainable lifestyle can feel overwhelming at first, but don't worry!

If you're not ready to completely switch up your habits, why not take a hybrid approach and incorporate small changes into your life gradually?

For instance, pick two days in the week you drive or get picked up in a car and walk or cycle the route instead.

Setting yourself small goals will lead to big positive changes in the long run.

Chapter 9

Action & Education

Study for solutions

What's the most powerful force on the planet, second only to your Dad's sneeze? Education.

Educating ourselves and others on planet-saving solutions is the most impactful tool we have for averting a climate disaster.

So start a free online course, take out some eco-books from your library or listen to a sustainability podcast and spread that knowledge!

Make your school & workplace plastic-free

You're now well on your way to becoming plastic-free yourself, but what can be done in your community?

The amount of plastic we produce each year weighs about the same as the entirety of the human race, and around 91% of it is not recycled.

Get a group together and fight for a plastic-free school, college or workplace. You have more of a voice and influence than you'd think.

Check out the Plastic Free Schools initiative run by Surfers Against Sewage for more resources.

Feed your local community

When it comes to saving the planet, you might not have money to give or the power to make policy in parliament, but what you do have is your time and energy.

So, why not donate some of that by volunteering at a local food bank, distributing surplus supermarket supplies to people who need them?

It's a great way to reduce food waste and help people in your community who need some support (and a friendly face).

Placard for the planet

Sometimes, sitting at home on our phones scrolling through negative climate news can make us feel powerless. But you're not alone.

There are people who care about our world as much as you do, and there's a chance they might live locally. Search 'local environmental groups' and 'local volunteer groups' online to find out what your nearest groups are.

Whilst social media can be great, it's no substitute for meeting people IRL and making actual plans – from marching and demonstrating, to lobbying governments, to cleaning up your neighbourhood parks and rivers, it's time to get out there and get active!

Get political

If you're too young to vote, it can be frustrating – you have a view but not a voice.

Fortunately, there are plenty of other ways to get your opinions heard when it comes to saving the planet.

Write to your local politician and hold them to account on the climate crisis.

Sign and share petitions and join protests for causes that you care about to keep the momentum going.

And remember – politicians work for you, and it won't be long 'til you'll be voting them in... or out! (Speaking of which, make sure you're registered to vote if you're old enough!)

Influence your inner circle

You've made your voice heard with your local politician and your school, but it's just as (or even more) important to influence those close to you – even your stubborn best friend and Dad!

Lead by example by showing off the things you've learnt from this book. Want your friends and family to start eating plant-based more often? Cook them some delicious vegan dinners and be prepared for your Dad to try to nick your last meat-free sausage!

You may think some people are beyond influencing, but it's possible and it's happened before!

Become a bookworm

Books are great (especially this one!), and you know where you can find a whole lot of them? Your local library!

They're completely free to use, and as well as having tonnes of books, they're an amazing resource for audiobooks, educational materials, newspapers and magazines.

So whether you need a book for school, for fun or to help you save the planet, libraries offer a great and sustainable way to access a wealth of valuable eco-knowledge.

Improve your local area

Ever thought your local area is lacking in greenery or looks a bit messy? Then do something about it!

Search 'local conservation group' or 'support local community' online and get involved with cleaning up your neighbourhood parks and rivers, planting trees or picking up litter.

You'll meet like-minded people who care about the planet, help support your local community AND learn some new skills. Win. Win. And win again.

Vote green

One of the best ways you can help shape the global conversation around climate and the environment is to support your national Green Party.

Green politics is a growing force around the world, and the more of us who get behind them, the louder our collective voice will be when it comes to saving the planet.

You can look up your regional Green Party online and find out how to get involved – most will already have local Young Greens groups where you can get together with other young people who care about the same issues.

Spare some (climate) change

When it comes to saving the planet, every penny counts!

We know life is expensive, and having money to spare is a rare and beautiful thing, but any time you find yourself with some loose change, why not put it into a 'piggy bank' (real or virtual, either works)?

You'll be surprised at how quickly all those little bits and pieces can add up, and once you've saved enough, you can do your bit for the environment by donating it to an eco-charity. Thrifty and nifty!

Look after your environ-mental health

Some days you'll feel like you can save the world single-handed, and others you'll feel exhausted at the job ahead.

Eco-anxiety is just as real as other forms of anxiety and can have just as big an impact on your overall mental health. So be gentle with yourself.

Now that you've taken to the streets, schools and your local community to save the planet, look inward. Mindfulness can be a great practice to help you navigate anxious thoughts and make clear choices about how you can make a difference.

Stay present, stay positive, keep learning, keep active – together we CAN save the world!

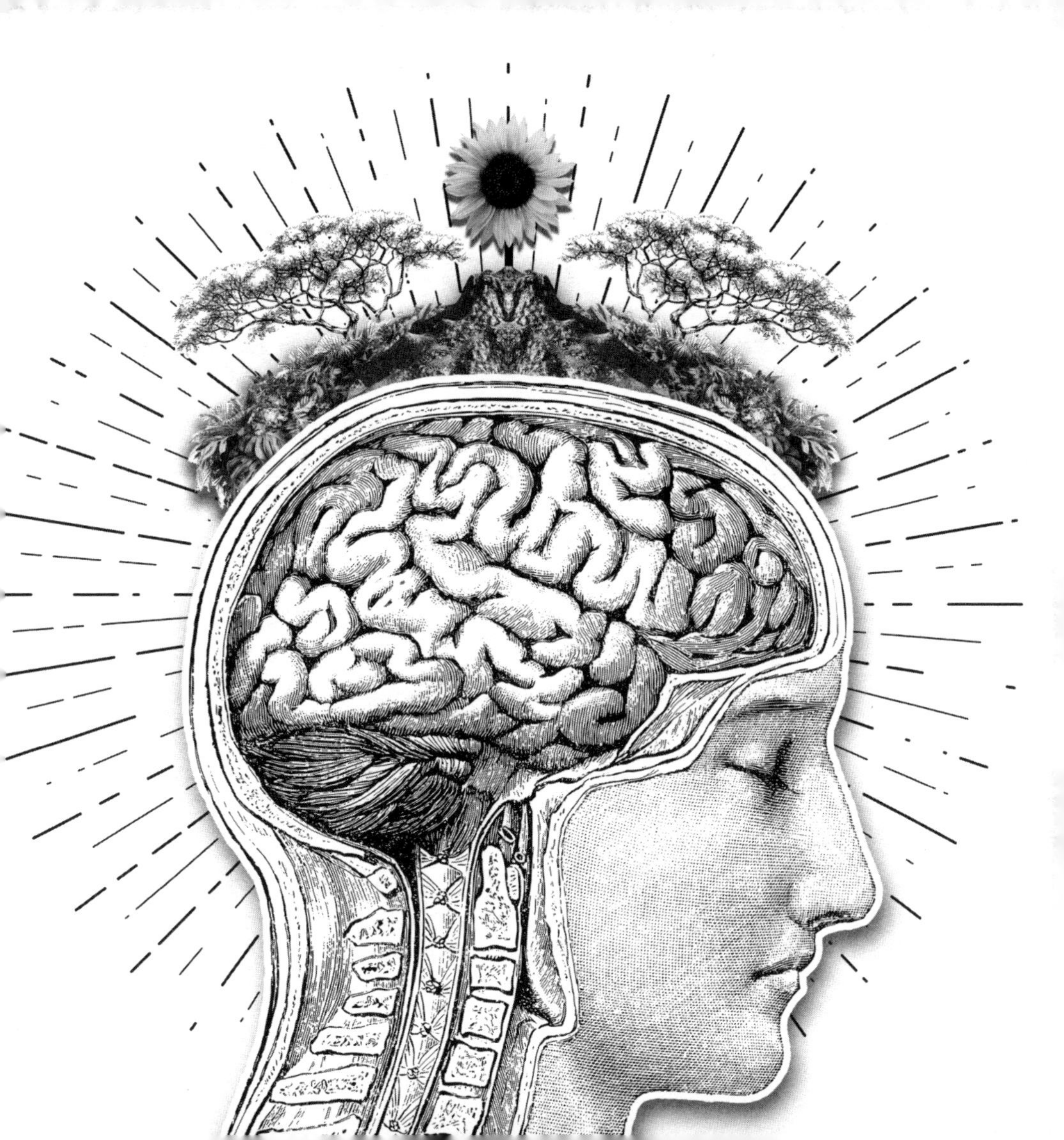

Join an online eco-community

We already know what you're thinking...
WHERE can I find an online eco-community that will show me how to action all these tips and outline the latest climate news?!

**Cue shameless plug*
We just so happen to already have one set up.

Join us by searching **Earthtopia** on TikTok and Instagram to learn more about how we can save the planet together!

10 Point Eco Action Plan

Now you've read the tips, here's an action plan to help you take the steps towards living in harmony with the planet:

1. ☐ **Getting Started** – Speak to a friend, family member or teacher about the climate crisis and how you feel about it
2. ☐ **Fashion** – Save one piece of clothing from the bin, by repairing it, donating it or selling it
3. ☐ **Food** – Go meat or dairy free for a whole day
4. ☐ **Home** – Repurpose one item that was destined for the bin into something new
5. ☐ **Online & Tech** – Donate an old mobile phone that's been sitting in a drawer

6. ☐ **Shopping** – Switch your shampoo, shower gel or hand soap to a natural alternative
7. ☐ **Nature** – Pick up one piece of litter each day for a week
8. ☐ **Travel** – Spend a day walking, cycling or getting public transport instead of driving
9. ☐ **Action & Education** – Volunteer a few hours of your time to a local environmental group
10. ☐ Pass on your experience and knowledge to others, to create a ripple of change!

Further Resources

Websites

Environmental Facts

CitizenSustainable.com

Climate.mit.edu

Curious.earth

EarthDay.org

FriendsOfTheEarth.uk

Greenpeace.org

SoilAssociation.org

UN.org/ClimateChange

WWF.org

Wrap.ngo

YPTE.org.uk

Eco Hacks

AlmostZeroWaste.com

ColdWaterSaves.org

EasyEcoTips.com

EnergySavingTrust.org.uk

GreenMatters.com

Moralfibres.co.uk

Good News

Euronews.com/green

Positive.News

ReasonsToBeCheerful.world

Green Money

Bank.green

MakeMyMoneyMatter.co.uk

Wildlife

DiscoverWildlife.com

Msc.org

WildlifeWatch.org.uk

WorldAnimalProtection.org.in

YPTE.org.uk

Books

Braiding Sweetgrass by Robin Wall Kimmerer

Climate Justice by Mary Robinson

Climate Resilience by Kylie Flanagan

Earthshot: How To Save Our Planet by Colin Butfield and Jonnie Hughes

How Bad Are Bananas? by Mike Berners-Lee

Is it Really Green? by Georgina Wilson-Powell

No One is Too Small to Make A Difference by Greta Thunberg

Not the End of the World by Hannah Ritchie

Ravenous by Henry Dimbleby

The Book of Hope by Jane Goodall and Douglas Abrams

The Climate Book, edited by Greta Thunberg

We Are the Weather by Jonathan Safran Foer

Follow Earthtopia

Instagram: @earthtopiauk
TikTok: earthtopia

This book was produced by 33Seconds, an
award-winning communications agency specialising
in climate, technology & lifestyle
www.33seconds.co